Animals in Deserts

Julie Murray

abdobooks.com

Published by Abdo Kids, a division of ABDO, P.O. Box 398166, Minneapolis, Minnesota 55439.
Copyright © 2021 by Abdo Consulting Group, Inc. International copyrights reserved in all countries.
No part of this book may be reproduced in any form without written permission from the publisher.
Abdo Kids Junior™ is a trademark and logo of Abdo Kids.

Printed in China

052020

092020

THIS BOOK CONTAINS
RECYCLED MATERIALS

Photo Credits: iStock, Shutterstock

Production Contributors: Teddy Borth, Jennie Forsberg, Grace Hansen

Design Contributors: Candice Keimig, Pakou Moua, Dorothy Toth

Library of Congress Control Number: 2019955582

Publisher's Cataloging-in-Publication Data

Names: Murray, Julie, author.

Title: Animals in deserts / by Julie Murray

Description: Minneapolis, Minnesota : Abdo Kids, 2021 | Series: Animal habitats | Includes online resources
and index.

Identifiers: ISBN 9781098202088 (lib. bdg.) | ISBN 9781098203061 (ebook) | ISBN 9781098203559
(Read-to-Me ebook)

Subjects: LCSH: Animals--Habitations--Juvenile literature. | Habitat (Ecology)--Juvenile literature. |
Desert animals--Juvenile literature. | Deserts--Juvenile literature. | Desert animals--Behavior--Juvenile
literature.

Classification: DDC 591.52--dc23

Table of Contents

Animals in Deserts

Many animals live in
the desert.

Jack rabbits have big ears.

These keep them cool.

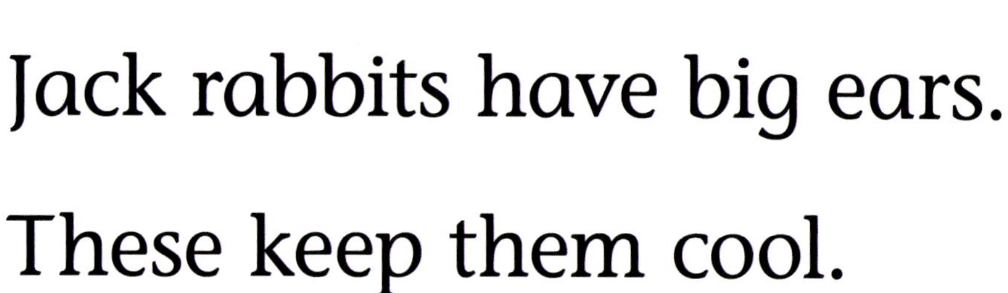

Lizards live in the ground.

This keeps them cool.

9

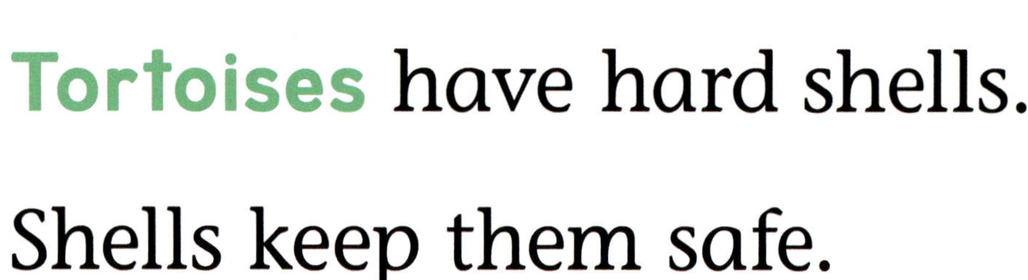

Tortoises have hard shells.

Shells keep them safe.

10

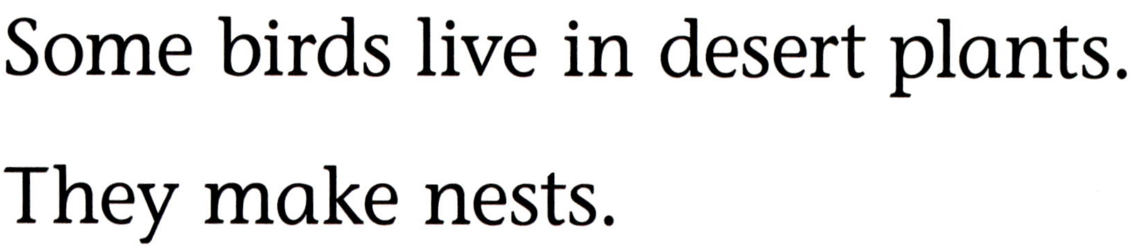

Some birds live in desert plants.

They make nests.

13

Some birds like to eat
desert plants.

Costa's hummingbird

Watch out! Some snakes
are **dangerous**.

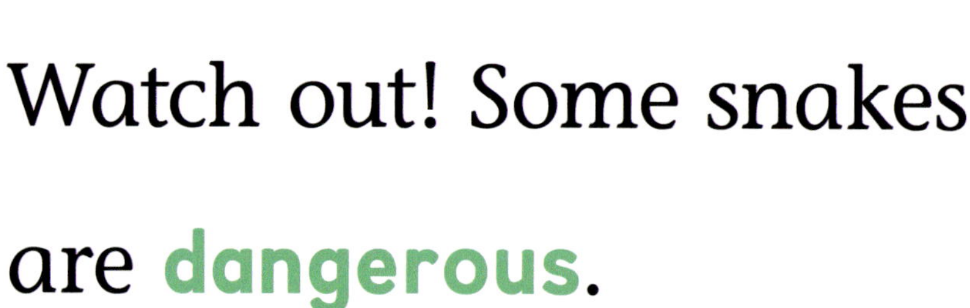

Great Basin rattlesnake

Owls live in plants.

They hunt at night.

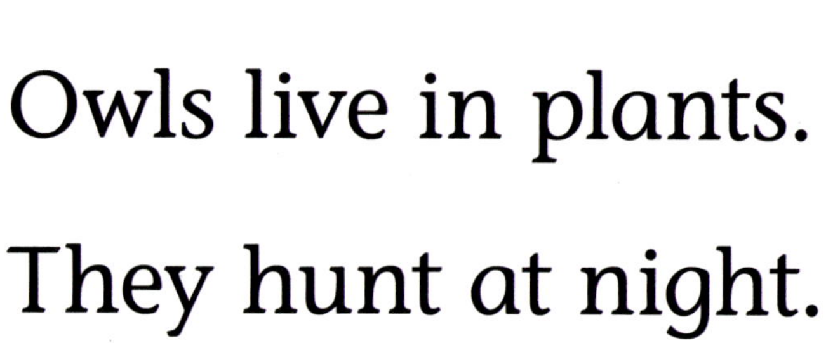

great horned owl

Roadrunners are fast. They can run 20 mph (32 km/h)!

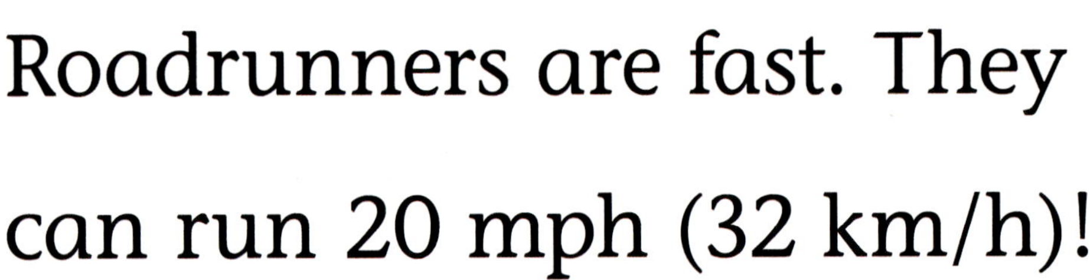

21

More Animals in Deserts

Cape fox

javelina

mule deer

tarantula

Glossary

dangerous
likely to cause harm. Not safe.

tortoise
a turtle that lives on land.

Index